A CANAL PEOPLE

THE PHOTOGRAPHS OF ROBERT LONGDEN

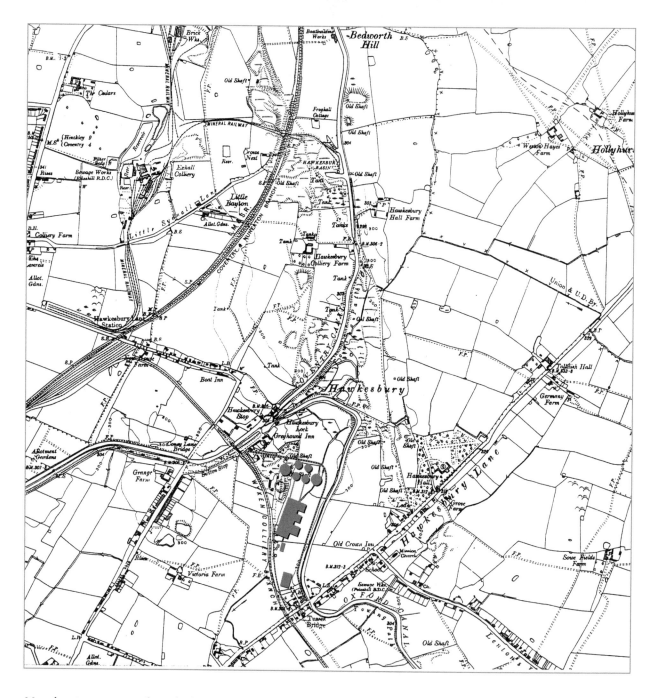

Map showing position of Hawkesbury Stop or Suttons, *c.* 1920, with the Coventry power station superimposed.

A Canal People

The Photographs of Robert Longden

Sonia Rolt

The History Press

For George and Anne

First published in 1997
Paperback edition first published in 1998
Reprinted in 2001, 2004

This edition first published in 2009
Reprinted 2013, 2017

In association with British Waterways and The Inland Waterways Association

The History Press
The Mill, Brimscombe Port
Stroud, Gloucestershire, GL5 2QG
www.thehistorypress.co.uk

British Library Cataloguing in Publication Data.
A catalogue record for this book is available from the British Library.

ISBN 978 0 7524 5110 7

Typesetting and origination by The History Press

CONTENTS

The man behind the lens:
Robert Longden, a self-portrait.

INTRODUCTION

Ihave been haunted by these pictures for almost five decades, and not only because I was involved in the life that is depicted in them. I hope here to be able to tell something of the story of this haunting and of its resolution.

Most people know by now that a government scheme in the last war brought young volunteer women to help work the family narrow boats conveying goods throughout the canal system. It had taken some time to bring into being but by 1942 the trainer Kitty Gayford, who had come onto the boats herself in 1941, was training teams of girls, three at a time, to a pair of boats, motor and butty. This was for the Grand Union Canal Carrying Company, Southall, a large firm of canal carriers based at Bulls Bridge, and the whole scheme operated under the control of the Ministry of War Transport. Thus, a number of women from every variety of background had their first encounter with the boating community. Most of the trainees, as they were called, left immediately the war ended. They had reacted to this encounter in various ways and with varying degrees of skill in the work and, considering how few of them there were, a surprising number recorded their experiences in one way or another. Most of those writings are mentioned in the pages that follow.

Many found the experience both astonishing and enriching. The boat community – the descendants of those families recorded during the last half of the nineteenth century and perhaps earlier – had, in their lives and the structure of their community, some elements of pre-industrial, even pre-enclosure, village society. This was evidenced still in their sense of cohesion yet independence, in their self-reliance, strong loyalties and strict family morality.

I did not leave. I ventured in marriage into the community and my life as a working boatwoman continued for another five years from this time. My friendships with boaters were naturally extended and my respect for them only increased. I never doubted that the virtues I had perceived and about which society then, and even more now, so bewails in their loss, represented a separate world of which I was privileged to become a part and to which I had committed myself. Of course, it did not seem in the least like that at the time. Inclination, a careless confidence and perhaps a deep hunger for community and family, which a dispersing and itinerant colonial background had done nothing to satisfy, were more likely factors.

A benign and patriarchal agent and firm of coal carriers – Samuel Barlow – was behind these years, bringing our working boats frequently to Hawkesbury Junction – Sutton Stop or Suttons, as it was known to the boaters and locals. Here the Oxford and Coventry Canals meet and here the distribution of orders for loading coal and the general organization of the boating life took place.

Towards the end of the 1940s I became aware at Suttons of someone, a photographer, who was often on hand, minutely observing the boats, the boaters and their movements in this setting of alternating motion, activity and rest. Gradually a friendship was made with this remarkable and self-effacing man. His name was Robert Longden and that was all at first I knew of him. He would come to the boats for a talk on our successive visits to this same place and, from time to time, he gave me a few of the photographs he had taken. These gave me enormous pleasure. I could see that by his presence there, his skill as a photographer and his patient concentration on this one place, so many of the activities which made up the whole life in all its variety could be caught by his camera. He knew my enthusiasm and gave me more of his work. This precious and quite slender collection came with me when I left the canals.

A new life and other involvements intervened when, with only few references to the canal life and the changes to it, his photographs might be looked at again. It was as much as twenty-five years later that I could yield to the pressure to seek out what remained and yield also to the urge to find out what had happened to the photographer and his pictures. I knew, and had always known, that they were important and there seemed to be no news in the canal world of either him or them.

Without the help of Mr Bert Dunkley of Coventry, with whom I had never quite lost touch, I doubt I would have succeeded. He was able, after taking some trouble, to give me a crucial address, that of the son of the photographer. With apprehension I learned the news that some of the photographs might be damaged or lost, some not even survive.

I went to Coventry to visit the son, also called Robert, and his wife Audrey. They were kindness itself, but had to tell me that when Robert Longden died in 1957, aged seventy-eight, he left a widow who, uninterested in his photographic work, had thrown away his photographs and the negatives he had left and given away his cameras (Leicas) and equipment. This came as devastating news to me. However, there was something else that had, apparently, survived. From an outside shed, his son brought in two small heavy boxes. I began to look at what they contained and lift the contents out carefully, and found with increasing excitement, that although spotted with damp and partly lying in mould within the boxes, there were well over a hundred $3\frac{1}{4}$-inch square positive glass slides. Most of them were of canal scenes, some were duplicates, and some were familiar to me from the small collection of photographs I owned. Exhilarated and not quite certain what might be the next step, I was allowed to take them away.

I started by asking for advice from my friend the photographer Eric de Maré, who had himself taken some of the classic images of the canals. He would know what I should do. He gave me a friend's advice and sent me to Messrs Roy Reemer in Fleet Street. Here a Mr Savage worked on the slides to clean and reclaim them from damage. He also made me a set of negatives and printed them all to full plate size. They began now in this form to come back to me and before long all were done – collectively amazing and stunning in their impact. Although anyone experienced in photography would have realised that they were several removes from Longden's originals (in the printing of the time, the 1970s), they revealed themselves as wonderful grainy, deep velvety blacks and whites, far indeed from the subtle range of tones and delicate greys in the original photographs I owned. Nevertheless, here was the same message, in another and very strong form. They were now to me as satisfactory and safe as I had been able to make them.

I wanted Longden to be recognized as one of the great canal photographers, perhaps indeed as a great photographer without category. To forward this aim and try to make a book from them I went to the publishing house I regarded as the best but, in the kindest way, I was turned down. After this, I have to confess, I let the matter rest. They were safe, the photographs and the set of negatives existed, and the slides had been returned to the family, who were now aware of their significance. It is the passage of the years alone which has made me feel some kind of duty requires a last effort. A sympathetic, even enthusiastic publisher, contemporary improvements in what revival of old images is possible, could make them a more common currency in their revelation of what the boating life was truly like.

It is also my hope that, before the last of those who experienced narrow boat trading on the canals in its heyday and its decline leave the scene, they may be prompted by these photographs of so much that they knew, to remember more themselves. Persuaded by their children or grandchildren, or coaxed by some of those who in increasing numbers are becoming more attentive to the boating life, perhaps interest and understanding can move a little further into a life that every day becomes history. This is the story of the pictures.

What do we know of the man who took them? Alas, very little it seems. I liked him so much. He remained around at Suttons into the very early 1950s as I knew, but I would like to know when he finally stopped bicycling to Suttons, for how many years in all he came there. Was it just the three-and-a-half years I can identify? Was he discouraged as the fine heyday of postwar traffic faded with the cessations of trade and lost contracts? Since I found the slides and first looked at them I have always wondered how he used them, how he might have spoken when showing them. Even with the most careful handling and consideration when I first took them out of the boxes, I could perceive no order or visible meaning in their arrangement, though perhaps, once upon a time, there had been such an order, now lost. His family, at all stages, could tell me very little about him and his life in detail, though I knew well his appearance. He was small, fairly reticent – until in conversation with sympathetic listeners – and neat, in my day always appearing at Suttons from his bicycle, walking about or resting by a bridge, with bike-clips still on, usually a mackintosh on or over his arm, frequently a black beret on his head. He certainly communicated with others besides myself and distributed other photographs to those he knew who had appeared in them.

His life had been spent in the machine tools trade. He was first apprenticed and finally became, as I knew him, a master pattern tool-maker, only recently retired from the firm of Alfred Herbert in Coventry, with whom he had passed the whole of his professional life. I was told he always ate with a set of small cutlery which he had made himself. He was domestically silent, even rather morose. The somewhat sad regard of the surviving self-portrait seems to show both professionalism and the presence of someone who knew completely what he was about; who might not suffer fools gladly. He became President of the Coventry Amateur Photographic Society and exhibited in their shows. I saw a few of such remaining works, but not many. While technically proficient they had none of the force of life and action present in the canal collection. For the time they are what would be expected: controlled soft focus or enhanced definitions, sometimes sentimental in his interpretation. It is easy to see his inclination struggling to pull this canal subject towards the same outcome: serious-faced infants struggling with artefacts too big for them under the proud half-smiling gaze of a young father. In such a scene, nevertheless, lies all which made the community remain beyond the range of serious trade union interventions – though it was tried once or twice over the period of the life of the domestic narrow boats.

The subject seems to have been too strong for Longden to soften it. We can only suppose that, after a long life of work, at last he had the time to pursue his hobby expansively. For these few brief years and in these new images of an old life, the man and his subject finally met.

In the pages that follow I have used the best of the slide collection and the greater part of it, only adding one or two of my own small photographs which fill out the interest of what is going on. I had first wanted the images to appear with *no* comment between them and the observer, using some form of notes at the end of the book. After much discussion I succumbed to the need for some aspects of the activities shown to be explained and that much more of the life could be understood if they were captioned as they now are. I hope that any faulty interpretations I have given will be immediately challenged and that such a method will not in any way lessen their impact.

So much more could be said, but in the limitations of the book format I must mention something more. In the background of many of the pictures lies the insignificant and transient structure known grandiloquently as the Boat Control Office, or, by the boaters, as 'Mr Veater's shed'. It was more robust than it looks: double-skinned timber, with faded blue and white paint. Inside it felt more solid, like a Norwegian garden building. It was, in business hours, the domain of Mr E.J. Veater and Miss Edwards and it would be difficult to overstate its importance to the boaters of this time.

When I knew him, Mr Veater had already spent many years in the service of the Grand Union Canal Carrying Company. He spent in all twenty-six years with them before retiring in 1950. Because he was trusted, on him had fallen the onus of retaining the crews and families of those owners who worked boats always known as 'Number Ones'. The carrying company, in obtaining long-term contracts at favourable rates, had the effect of dispossessing many of these. Some of them retired, not wishing to join a uniform company of many boats in their livery of standard red, white and blue and subject to company rules. Some did join, recruited perhaps by Mr Veater, and transferred their skills and pride in the job to the new company. Some carried on, sub-contracted to the smaller Midlands companies which they knew and which could be counted on to continue the decorative and docking traditions they liked and were used to; and some finally joined these firms.

To recruit the family crews needed to man the enlarging Company fleet, which in turn was later wanted by the Ministry to handle the increased wartime traffic, Mr Veater travelled widely, calling at many centres of boating life and activity throughout the Midlands. This gave him an unparalleled knowledge of how the boating life worked and an understanding of the boating families and their needs. The inevitably piecemeal system of enterprising individual steerers finding contracts where they could, gave way to an overall control operated from the Boat Control Office. Here the need for coal at the London mills and factories was matched to the supplies available for loading at the Midlands pit-heads. The steerer of boats arriving empty at Suttons would go to the office for orders to load. The 'office', or the policy operating from it, could be flexible and correlate commercial need to family requirements to some degree. In the case of illness, death or childbirth, the steerer could usually be found short runs while the domestic butty could remain tied up, available for doctors, nurses or occasionally health inspectors, or later the wonderful Major and Mrs Fielding of the Salvation Army, whose boat *Salvo* was moored nearby. Mr Veater presided over all this, occasionally also arranging marriages, christenings and churchings with the local vicar at Longford. This many-sided activity made him similar to a good purser on a ship and, indeed, I always thought he did resemble such a figure and that the operation of the boating life was like a large vessel of mixed cargo, the passage of which nothing must be allowed to hinder. This 'vessel' certainly included trade and commerce and communication by water, the very purpose of this great underlying network built to serve it. I do not suppose that the idea of a main *pleasure* – let alone, as we must now say, *leisure* – use ever crossed the mind of Mr Veater or of his mentor, John Miller. The latter, chairman to the Company, wrote and spoke eloquently and convincingly of the postwar view ahead for commercial carrying.

Of course, nothing does quite end. Many more activities were woven into the gradual demise of commercial carrying. Even now, narrow boats still exist, authentic ones, moving about like ghostly presences in the hands of their new owners, occasionally indeed carrying goods and themselves the subject of historical interest and preservation.

What have the boaters themselves said of the endings and leavings of such a comprehensively harsh but beautiful life? Nothing much, it seemed, to start with, though latterly rather more, since interest does not grow less and those who survive find themselves regarded as icons or even phenomena – like survivors from another land. I think Charlie Carter was right when, listening to a recently taped interview, I heard him say of the boating life: 'We were free, see; we were free'; but I would want to contrast this with the reply I awaited from my dear friend Henry Monk, sitting some years ago in the still air and sunshine of my Gloucestershire garden. A very long pause ensued, then: 'It was all work, Sonnie, it was all work.' Perhaps it was, both freedom *and* work: a rare combination. But then Henry and his hard-working wife, Ruth, *did* take every opportunity they could, when away from their coffee-shop at Harpenden, to travel again the canals he knew – trying perhaps (who knows?) to recapture that still disclosing view, that as yet unseen vista.

THE PLACE

Apart from the places shown in the chapter 'Other People, Other Places', Robert Longden took these photographs during his visits to Hawkesbury or to the power station or any one of the colliery loading points and sidings. Here he might have expected to find the subject he was seeking, the boat families, their boats and the activities connected with them. All were within easy reach of the heart of the matter, the two basins of the Coventry and Oxford canals at Hawkesbury or Sutton Stop.

When John Sinclair, engineer to the Coventry Canal Company, carried out improvements in 1837, the two basins and the Coventry and Oxford canals were connected. A spectacular iron bridge installed at the same date carried the tow-path up and over the neck of water which joined them, from the Oxford canal to Longford and on to Coventry basin in the city.

A pleasing group of attendant buildings lay around the basins. The engine-house, chimney and engineman's house in darkest engineering brick, the soft tones of the eighteenth-century brick-built cottages and stables, and the rendered façade of the Greyhound Inn made up the whole, against which the boats and people passed to and fro.

The isolated canal settlement lay a few miles to the north-east of Coventry city centre. Barely more than a quarter of a mile away along the Oxford canal, and much less across the fields, loomed the power station. Surrounding these two contrasting settings lay abandoned ground, the legacy of early mining's old tips and shafts and disused canal arms to former colliery sidings. This made up a rural setting of rough grass, nooks and corners, where gipsies hobbled their skewbald horses and came sometimes with vans which rivalled new-docked boats in their decorations. Small brick pubs along the canal side and the occasional isolated cottage might have housed characters out of George Eliot.

Since the power station was demolished and the site cleared in the late 1970s, it might seem impossible to imagine the sheer mass of the building and the activity in and around it as it was then. From Hawkesbury you saw the bulk of the huge turbine houses, the chimneys breathing black smoke and the cooling towers leaching their slow clouds of white vapour.

This then was the background against which Robert Longden took his photographs. There have been many changes as a result of the cessation of trade and coal-carrying in the late 1960s, but the ingredients making up Hawkesbury are much as they were then. The news this year of a £54 million development plan which will bring housing, golf course, industry park and water attractions is change indeed! Fortunately Sinclair's Britannia Bridge, which looks in good heart, is earmarked for attention and care.

This scene is largely as created by John Sinclair's improvements in 1837. The engine-house, chimney and engineman's house were built when the second, and larger, beam engine was installed at the time. This was capable of clearing the drained water from Exhall Colliery and Bedworth Field, pumping it from a well back into the canal. Sarah Collins steers the loaded butty under the cast-iron roving bridge and through the Coventry and Oxford canal link.

When the link was made, the junction created those difficulties in manipulating the boats which gave the photographer his dramatic material. A loaded Grand Union butty is brought round the acute turn between the Coventry and Oxford canals. Mr Grantham holds the fore-end with a line from the bridge.

A fine view of the cast-iron bridge made at the Britannia Foundry near Derby, for the improvements of 1837. It proudly bears the makers' name, that of the engineer and the date. A view south-west towards Longford is seen behind an empty L.B. Faulkner boat.

The dramatic soffit of the bridge divides the views of both the Coventry canal stop-place and the Oxford lock. The bridge had an innovatory roller to ease the taut ropes, but this became stuck and the ropes ate into the stone pillar of the abutment.

Quartering the scene and moving east from the Oxford lock, the basin with its fuelling point outside the Greyhound Inn is now visible. In it lie two pairs of boats waiting for orders.

The small-scale domestic façade of the Greyhound Inn beside the basin makes a background for the fore-ends of two empty pairs. Near the shore lie a pair of Grand Unions with standard electrical headlamps and, nearer the camera, a pair of Fellows, Morton & Clayton boats. Note the fore-cabin on the butty registered for domestic occupation; this usually meant children.

Mrs Berrill meets Mrs Carter outside the door which led directly to the all-important shop. The projecting windows of the bar at the inn are either side of the door. Atkinson's Aston Ale was provided there and Rowena Nelson both served it, or served in the shop.

A peaceful scene reflected in the waters outside the Greyhound Inn. It is unlikely that these people have come from a boat, although they may have done. Some boaters had family connections at Suttons or owned cottages there.

This scene shows the great contrasts that existed in such a small area. Not much more than a quarter of a mile away from Hawkesbury stood the massive block of Longford power station, always known as 'Coventry Light'. Children, a rudimentary swing made for them and motor and butty cabins adjoining may mean a maternity stop. The children are thought to be Jeannie and Jenny Humphries.

There were usually boats on the blackened and rubbish-strewn waters underneath the cranes, gantries and conveyor belts which took the coal from them to the furnaces or to a coal stack. An empty boat lies along the more distant wharf.

A Götterdämmerung shot of the cooling towers, chimneys and a pylon at Coventry power station, all working at full belt. Crests of white steam filter into the threatening sky.

Domestic life carries on in this setting, as with the earlier picture of the children on the swing, and there may be a similar reason for its presence here at Tusses Bridge with the Elephant and Castle pub nearby. The wind drying the washing has been judged finely, and the steam and smuts from the power station are being dropped on the brick-built group at Hawkesbury a few hundred yards away.

An unusual shot showing a little used tow-path. The laden telegraph poles signify that a simplified way-leave has been obtained from the canal company, instead of from numerous landowners. The number of wires suggest trunk lines coming into Coventry.

Opposite the coal hoppers by the gasworks and near the Longford sidings where loaded trucks from the collieries were marshalled and shunted, a pair of quiet fishermen wait for a coarse bite.

The gasworks and their coking ovens, and a foundry just out of the picture, filled the air with unique and unmistakable sounds of metallic roarings and groanings. Two pairs of Grand Union boats are loaded and ready to leave.

For a long time I thought this was an aerial photograph, but Longden had taken it from the tallest of the Longford gasometers which was not used during the war as it was too conspicuous. From its height he looks back to Suttons and the power station, and on into the Warwickshire countryside through which the canal passes. I am reminded of Henry Mayhew in the balloon *Nassau*, seeing from above London the whole of what he had previously recorded so intimately.

THE PEOPLE

It has been asked, why the title *A Canal People*? After some thought it seemed best, to me, to insist on the indefinite article. (I had really wanted it to be called *A People* but the word 'canal' had to be included.) It was because Robert Longden, living in Fisher Road, Longford, and within easy bicycling distance of Suttons, concentrated his camera on the boat families congregating there or working their boats round the turn or through the Oxford lock. His pictures were of those who, *at that time*, worked almost exclusively for the firms, and their loads and journeys which brought their lives before him.

There were many other boat people elsewhere at this time and many related to those we see. These others were somewhere else on the whole system as it existed for narrow boat carrying. They may have been anywhere between the Mersey and the Midlands, congregating not at Suttons but at Middlewich or other gathering points, but here before us are those particular canal people whom Longden could observe and record on his visits.

There are not so many names involved and I shall list only their surnames as they are familiar to me. Most appear in the pictures or are related to those who do. Others, due to the exigencies of working, loading orders or docking, can be assumed to be near at hand, about to arrive or, sadly, gone off ahead the day before, by their absence off picture, but only just. Here before us are members of the families of Hone, Hough, Hambridge, Boswell, Nixon, Beechey, Humphries, Littlemore, Wilson, Carter, Ward, Grantham, Peasland, Bray, Lane,

Higgins, Gibbons, Lapworth, Griffiths, Skinner, Tonks, Wenlock, Whitlock, Taylor, Smith, Collins, Stokes, Townsend, Barrett, Berridge and Monk.

The people, about whom so much could be said, here speak for themselves through their presence, style and handling of the activities Longden puts before us. Since these days shown here, most existing books relating to their lives have been published and a number of films, broadcast programmes and, latterly, tapes have been made, which can be sought out. It seems to me that Robert Longden approaches nearer the people than anything else we have seen in both evocation and record.

Later on in the book I list other writings which, not restricted to just what must be said about Longden's photographs (the purpose of this volume), expand much more fully on the nature of both the people and their work. But, between the dry listings of registrations, occupancies of boats, the families in them and the tonnages they carried – and the impressionistic and heart-felt evocation of a re-created life *Ramlin Rose* – lie many other writings. Perhaps the book we truly want is still to be written. It could be by a descendant of one of those families who show us their lives and selves in Longden's scenes.

Some of the Barrett children propped up comfortably against their canal boat. The family at this time worked for S.E. Barlow. With them are three friends from the Humphreys family.

The dancing trio of cousins approaching the Oxford lock is Lucy Humphries, Rene Carter and Georgie Humphries, known as 'Bucky'. They are pulling a boat to the lock alongside one of the few whitewashed houses in the group at Suttons, which was later demolished.

Longest lasting of all 'Number Ones' of the old school, Rose Skinner steers under the bridge in the family boat *Friendship*, loaded with coal for Oxford. The boat has been docked by Herbert Tooley's yard at Banbury; 'side boards' have ensured a good load and keep the boat in level trim.

The Stokes family – Arthur, Nellie and daughters Joyce and Sarah – rest on their boats, awaiting orders in the Greyhound basin. The boats *Mimas* and *Rae* were always spotless, and keeping them so seemed both a way of life and an art form.

The two girls and Arthur Stokes sitting on the motor in similar circumstances, this time around the corner on the Coventry canal. Longden, for obvious reasons, often photographed this family and their boats. Ovaltine boats were registered at Rickmansworth, near Walkers' yard where they were built.

Mrs Rose Bray fills her decorated water-can at the tap behind the Oxford canal ticket office. You had to keep the knurled grip turned on for the water to flow, which ensured that fresh water, that most precious commodity, would not be wasted.

The affinities between Mrs Rose Bray and her boat are caught by the photographer. The Brays worked for Harvey Taylor of Aylesbury, often carrying wheat out of Brentford to various destinations.

An authentic and traditional boating bonnet worn by Mrs Phil Ward. She was one of three sisters whose marriages meant they all worked for different companies. The wearing of the bonnet survived longest on the North Midland runs and this suggests she is with one of the pairs of Fellows, Morton & Clayton. They were taken over by British Waterways and brought to less familiar waters, giving Longden his opportunity.

A familiar, beautiful boating face. Mrs Mabel Wilson (*née* Nixon) married to Ernest Wilson. They worked for Samuel Barlow Coal Company and their boats were equally beautiful and sparkling, despite Mabel looking after her two small children.

Mr Henry Monk snr in the hatches of his butty *Ironside*. A well-known portrait of this patriarchal ex-'Number One'. The picture clearly shows the brass chimney-chain made up of a series of strap clips cut from army canteen and canvas bags found in the great rag heaps at the paper mills.

Rose Jackson (*née* Hone) and Joycie on the stern of *Cylgate*, steering through the Coventry canal stop-place on the way to load coal. *Cylgate* was one of the Hones' three 'Number One' boats and the tail flowing so richly from the ram's head once belonged to a family horse.

John Knill, later Sir John, was one of the few outsiders to start trading shortly before this time. His boats, *Columbia* and *Uranus*, were worked with a great sense of pride and determination. Here he talks to Alice, Mrs Henry Monk, who appears to have visited the Greyhound shop.

An elegant portrait of Mrs Sara Bray, the mother of Arthur who appears with Rose on their Harvey-Taylor boats in many pictures. She seems entirely at ease with the photographer.

Posed here on a British Waterways boat, of no great interest to the photographer, is a group consisting of Mr and Mrs Bill Taylor (she was formerly Mrs Ikey Slater) and Clara Newbold. The dog joins in for the portrait.

Here Doris Beechey (Mrs Bill Grantham) and sisters, and assorted children, make a group which would have appealed to Augustus John. In such family boats wash-day was every day, but a meeting and sitting together in the sun is a rare enjoyment, not to be missed.

Sarah Stokes in the stern hatch of the motor *Mimas*. The boats were docked at Braunston yard. The shadow of the 'elum' (or helm) falls across the snowy whiteness of the scrubbed counter boards. The edges were always protected by ash strips which could be, and here are, scrubbed as white as paper.

Samuel Barlow pair *Cairo* and *Warwick* waiting empty at Bedworth bridge to load coal brought from Newdigate Colliery to the Bedworth arm behind the camera. Sonia and George and Jim Smith look at some of Mr Longden's photographs. Such photographs as I possess were handed across in this pleasing way.

Mr and Mrs Jack Skinner snr worked for Samuel Barlow Coal Company. They were the parents of the Jack Skinner who married Rose Hone seen earlier. Note the ingenious gangplank, fixed up to aid getting on and off an empty butty high out of the water. Here they are in conversation with Rose, Mrs Charlie Carter.

A Samuel Barlow butty *Gertrude* docked at Nurser's yard, Braunston, and here seen empty at Suttons, supports a number of the young Barretts from Sid Barrett's family. From the left are John, Ted, Ken and Phoebe.

From the roving bridge Longden catches a skilful manipulation as the motor pulls the butty around. Steerer David Hambridge on motor boat *Gort* from Braunston. He was felt to be much favoured in the Barlow firm when he later moved to *Ian*, with a powerful new Gardner engine.

Mr and Mrs Sid Gibbons with Sid jnr, very much intrigued by the camera and sitting on the cabin top. He may be harnessed for safety. The breasted-up boats have come for orders and Mrs Sid (*née* Littlemore) leans on her tiller to help the turn that the boats are to make. Everything is *à point,* and all looks relaxed and cheerful.

THE WORK

The boats which *were* the work, although their presence is manifest in every shot, are not usually – except in rare cases – loaded, trimmed, cleaned up and gliding in a well-behaved manner from one disclosing canal vista to the next. This is what people think of as canal-boating and it was a great part of the work, as was indeed the other publicly seen activity, when boats were climbing or descending through locks.

In the majority of the pictures in this section the boats are at their most recalcitrant, in their unwieldy but powerful forms. They are being pulled and pushed to loading points or away from them, or being manipulated to pass under bridges and around tight turns. Finally, in many cases they are simply tied up, waiting for the action to begin. In the minds of all those shown, this loaded travelling is to come, the very purpose of the work being to load the commodity, in this case coal. Then it had to be got to the various factories, mills and power stations needing it. On one occasion only, so far as I know, Longden followed a loaded pair up Braunston locks and here, with longer captions, I have tried to explain this procedure for a two-handed pair in rising locks.

Work was hard and unremitting, but to move off with prepared boats and a load always brought a sense of satisfaction, a release and a feeling of freedom, with thoughts of the long journey, the pounds, the tunnels and locks, ascending and descending over the wide country to come. Every nook, every bend, every tree, every wharf and bridge would be known, but each time appeared anew. A litany of named locks had to be passed on the way, for instance, to the 'jam 'ole' (Kearley and Tongue) at Southall – the jam factory, greedy for the coal thus brought. This then was the geography of the work.

With barely a change, except for the season, the same family boats, now loaded, are hauled cheerfully towards the camera and into the Oxford lock. Mrs Sid Gibbons is with daughter May and Georgie Hambridge: they were later married. On the left is little Sid who, in the last plate, sat on the cabin top in more congenial and summery conditions.

In this and the next plate a key figure in the proceedings is seen, Mr E. J. Veater, Traffic Control Officer at Hawkesbury. He and Miss Edwards, his assistant, received the necessary information about any empty boats arriving and could give orders to steerers to proceed to the collieries. Here he talks with Mrs Slater, and the family dog, as they set out loaded with coal.

A Fellows, Morton & Clayton pair are now on the coal run after their boats were taken over by British Waterways when the firm ceased to trade. The steerer's wife has the clearance papers for the journey south with approximately fifty tons of coal. Mr Veater sees them on their way.

A dredger on maintenance work passes along the Oxford canal and under the gantries and continuous feed belts at the power station. Note that the dredger has separate pontoon sides which can be removed on narrower stretches of the cut or to pass through bridges. Samuel Barlow Coal Company motor *Daphne* rides high out of the water as the grab above has lifted the coal from her stern and transferred it to the hopper above the conveyor belts.

Some normally long-distance boats worked on short runs from one of the collieries to the power station if there was a shortage of Joey or day-boats, or if there were no long-distance coal orders for them. Some boats worked short hauls more regularly, among them the Barretts. Here on the left, Rene Carter cleans down the butty boat *Dorothy*.

A very well-known setting. Boats lying alongside the waggon siding in the Bedworth arm and being loaded with lump home coal from Newdigate Colliery, bound for Oxford or Rugby. This was a dusty job, requiring major cleaning down afterwards as the coal dust got into everything. Mr Lenny Gibbons, an ex-boater whose son was pictured earlier, worked with the trucks and the loading and is seen here walking away from the chute.

A cluster of boats waiting to load outside the same Bedworth arm. This is a mixture of Grand Union, Fellows, Morton & Clayton and Barlow boats. Tom Hambridge on *Hawk*, lying on the outside, and behind him his son entirely enveloped in smoke, stands on the fore-cabin of the butty. He has just lit a fire in the small fore-cabin stove. Rough sheeting up at the stern gives a dry area for children to play in.

The photographer is on Bedworth bridge and it is clear there is a friend on the bank to the right: 'I'll see y' back in the New Inn when we gets tied up at Pooley.' The boats may be going to load at Atherstone, Griff, Baddesley or Pooley, probably bound for 'the Light'. Caleb and Annie Lane are both on the motor of an S.E. Barlow pair.

In Griff Colliery basin small Woolwich Grand Union boats are loading. Had they been the bigger boats loading, all removable superstructure and objects would have been taken down to get under the very low bridge at the entry to the basin, which gave clearance sometimes as little as half an inch. In the hatches of the outside butty stands Mrs Clara Collins.

Back at Suttons, Arthur Bray, Charlie and young Roy Carter lead the empty boat into the Oxford lock. Note the capping to the wall, where the tow-path, having changed sides, descends from the roving bridge, a smooth surface for a horse-boat tow-line if needed.

A cheerful bunch helping out with the tow, not all of them identified, but Rene Carter leads the line.

A fine study of effective but relaxed physical effort. The pull is from the masthead.

The empty Joey or day-boats in the background are on the island at Pooley Hall. In the foreground, great physical effort is needed to shaft the fore-end of this loaded Grand Union Woolwich boat round. It is also possible to see here the traditional method of folding the front top-cloth and bringing the side cloths up to the cratch frame. These cloths covered the cargo when it needed protection.

Mrs Clara Humphries, sister to Jack Skinner snr, shafts the empty Barlow butty. The butty's position means it has come back stern first from 'the Light'. The boats must be on the short day runs described earlier.

The same occasion gives an excellent view of the Traffic Control Office, occupied by Mr Veater and Miss Edwards. This was the hub of the whole operation in dealing with the numerous boats that converged here.

A wonderful sequence showing the movement of a Joey or day-boat round the turn, in the skilled hands of Jack Lapworth working single-handed.

The camera shows just the right deployment of weight and effort needed to make the heavy boats obedient.

The boats would be tied up together fore-end to stern-end, but would have to work separately through the single lock. The small scale of the day-boat cabin and hatches is evident.

Jack Lapworth pulls Joey boat *Rose* back, to tie the two boats together before setting off with them behind the horse for 'the Light'.

An S.E. Barlow motor *Caen* fuels up at the Nelson's Gasoline tanks. Rose Whitlock contemplates.

The same scene, now including Bill Whitlock, shows Ron Wilson, also involved in the operation, looking out of the engine room. The Oxford canal office from which the 'gauging' was done – based on measures of dry inches down the boat's side, which determined the correct tolls to be paid – is visible in the background.

Bill Whitlock shafts the butty's stern to make the turn under the bridge. The cabin doors are shut because the Whitlock children, Joan and Michael, are below in the cabin. I like to think that the shadow of Longden on the bridge above shows faintly on the water or boat.

Mrs Edna Carter (*née* Tonks) pulls her empty Ovaltine butty *Hector* back to moor by the Traffic Control Office.

Butty boat *Hector* moors up. This picture shows well the shackle-pin set among the protective ash strips and decorations. The hemp checking-strap would be attached with a spliced loop and this strap trailed in the water when running in down-hill locks.

Pulling the butty up to the lock as the motor moves out. Longden had recognized the photogenic quality of Mrs Carter. It can be seen that she has had a hard time cleaning up after loading.

Mrs Jane Wilson holding an empty butty in the Greyhound basin.

Here she moors it temporarily from the mast. In the background is the Oxford canal toll office. Stop-planks for shutting off a portion of the canal were kept in the brick extension at the back of the office.

The Oxford canal toll office clerk talks on the bank while the fore-end of this carefully sheeted-up motor is pushed with the shaft by Tom Wilson into a good position for the turn. It is sheeted-up as if it is loaded with dry coal, possibly for Springwell or Coppermill lock.

The S.E. Barlow butty *Hardy* in the Greyhound basin. Notice the 'cobweb' fender to protect the 'elum' and the castle panel. This is very characteristic of the style of S.E. Barlow's boatyard at Tamworth. Mrs Jane Wilson leans between the doors of the butty. On the right, an L.B. Faulkner boat.

Bringing a loaded boat round the turn from the Coventry to the Oxford canal. The fore-end of the motor in steady ahead-gear will be held as the cotton line is played round the bollard, and the stern swings. Mrs Laura Carter controls this manoeuvre.

Sometimes the only way to get round is by muscle power and hard pushing on the long shaft. Young Abel Beauchamp holds the boat while his mother Edith puts all her weight on the shaft.

Arthur Stokes makes use of the filled Oxford lock to clean out his 'blades' (propeller) and drive-shaft. The sort of rubbish which would get caught up as the boat moved, lies in the foreground. Nellie and Sarah look on.

The S.E. Barlow butty *Victoria*, opposite Kelly's Timber Yard on the Coventry canal link between Hawkesbury and the city basin. Her steerer seems to be concerned with a leak which he is plugging in her stern planks. By letting the boat drift across he can reach the affected plank.

Mr Taylor worked Joey boats with his brother Bill and two sisters. The family were known collectively as 'Four Boat Joey' and worked short lengths from collieries to the power station.

Mr Griffiths shafts the tug *Enterprise* for the Warwickshire Coal Carrying Company, towing Joey boats on short runs. He and Mrs Griffiths lived at Griff and their two girls Valerie and Velma had been to school and were 'scholards'.

Here is Valerie, a scholar daughter, on one of the Warwickshire Coal Carrying Company Joey boats at the Coventry power station.

The same family. As I look at these two pictures I am persuaded that, on this rare occasion, father and daughter are posing for the photographer in this dramatic setting. The photographer seems to have got himself positioned on the fore-end of another boat. I never knew him do this anywhere else.

Sarah Anne Carter on the bow of the Harvey-Taylor butty with the cooling towers of Coventry power station in the background. Jim Peasland is walking along the top planks.

Ernie Grantham provides the photographer with a favourite subject, ropes and arms. Note the neatly spliced tail to the cotton line.

Shirley Peasland looks serious and efficient, tying the empty boats up short together. They are coming out of the lock on their way to load.

THE LIFE

It seems rather arbitrary to divide the 'life' from the 'work', since everything and every recollection given shows that they were one and indivisible. However, it is worth pointing out that at Sutton Stop, Robert Longden captured those times when the *life* of the boaters to some extent took precedence over the *work*.

Although the life portrayed here is now entirely ended, at the time very little effort was made to record it. Certainly none was made, except in very broad terms, to enter into the experience of the whole life of a boatman or a boatwoman. Those accounts by the wartime trainee boatwomen related rather more to the practice and system of boating – what you had to do – and to the subjective views of the effect of the life and the people on the writer.

One exception, for me, is the novel *Maidens' Trip* by Emma Smith, who was one of the very effective trainee boaters of the war. This beautiful, vivid re-creation met the whole thing head-on and made very little effort to explain or, worse still, intellectualize or comment on the survival – as survival it undoubtedly was – of this separated community. Besides her companions, their characters sparingly and penetratingly evoked, you see the boat people, you hear them speak, you smell the cut water and hear some of the noises (this last very important). None of these things can really be successfully evoked (as is now attempted) with engine and accordian background to audio tapes and with well-placed questioning of some of the survivors.

This can take place because the life is not quite over. There are still some boaters, and the children of others who have heard tales and who remain to speak to us. Half a century after *Maidens' Trip*, Sheila Stewart, not a boatwoman herself, explains the pressures which made her re-create, through years of research and friendship with survivors, in particular Ada Littlemore, her composite representative boatwoman 'Ramlin Rose'. This is unlikely to be superseded, for indeed Sheila was aware it was published in the nick of time. I have one comment only to make, other than praise for such a moving work. Because 'Ramlin Rose' is such a composite character, she endures altogether more of the tragedies and pains likely, as it seemed to me in my own limited experience of seven years, to come the way of a single individual boatwoman.

In that comparatively short time I knew of only one terrible and tragic accident. There were, however, a few near misses as children occasionally fell in and were fished out. I was impressed more by how agile and secure they were as they moved through the life. Of course, as in every life, family disasters bring grief and trauma, but a natural ebullience seemed to prevail, and humour and badinage must have aided recovery besides the close physical aid always present between the families.

Perhaps that word 'freedom' which crops up again and falls from the lips of those who, over the years, begin to say what they liked about the life has a lot to do with it. Somehow 'Ramlin Rose' seems more imprisoned than some of the survivors ever thought they were, wives as well as husbands, as they now recall the days and times they knew. Into such rare spaces as came their way rushed the activities of 'life' which 'work' had not allowed: washing and drying clothes, shopping, cooking, maintaining or repairing the boats, decorating, crochet for the cabins, the arts of life; or simply the opportunity to recover from the last trip from London to the north and the breakneck journey for orders to load again in the coal field.

The vital ingredient of daily life, washing and keeping clean, rarely found so good an opportunity as this. Space for a good line, a handy support for a plank table and a fire for boiling water for clothes (which must be off the picture) were the main ingredients. Josie Grantham (*née* Beechey) with her sister Netta, who married Henry Wilson, share the domestic labour.

Mrs Emma Humphries, on her Samuel Barlow Coal Company butty *Mafeking*, peels spuds while one of the twins looks on. Note the powerful shackle-hook which appears in close-up.

A moment of leisure while a big tank fills up with fuel from the Greyhound pumps. Samuel Barlow boats and a young George Humphreys in attendance.

Washing, shopping, cooking and visiting could all take place when a weekend without orders gave time for them. Jack Skinner leans on *Kent* in the foreground and Polly Safe is behind him in her butty *Ash*. The Blue and Yellow livery on her boat is brand new; it was not liked.

It is not known whose brilliant interior this is, but consensus says that it is that of Ron and Sarah Hambridge. It shows crochet, plates, brass rail and knobs, burnished stove and wedding group, plus newspaper to cover the perfect floor – all the ingredients of a well-furnished cabin.

This little Harrison girl, Dorothy or Anne, has assembled a remarkable collection of objects to play 'shops'.

A poignant picture of playtime which speaks for itself. Ernie Wilson's butty *Mary* displays the handsome Braunston yard finish of a recently docked boat and shows up the rough finish of the fore-end of the boat moored behind, urgently needing maintenance and 'docking'.

Martha Carter, with Jeannie helping, fills a collection of water-cans. She married George Humphries. The middle can is a fine example of the tin-smith's craft.

Charlie Carter in nonchalant style with a strong hand on the tap. The picture shows the now demolished lime-washed brick house in the background.

Shirley Peasland takes
care with the same task.
Sometimes, if a family was
a big one young members
could be spared to help with
the work on the boats of
close relatives.

Mud larks on a sunny day outside the Greyhound Inn. Joyce and Dennis Hambridge make the most of summer.

Handsome Phyllis Berrill with her babies at Suttons. The boats are loaded and ready for the run to the London area. The twins look very new and were probably born 'back of Mr Veater's sheds', where they were well looked after by a local doctor and midwife. In such circumstances Mr Veater would try and keep the steerer in work, using his motor on short runs to the power station.

Mrs Nellie Stokes (*née* Littlemore) and the burnished throne of her butty *Rae*. Everything is in place and everything is perfect. The motor with a long line over the paired single locks at Hillmorton will pull her gently in.

Mrs George Smith leaving the shop at the Greyhound after stocking up for the run down country to the paper mills at Croxley or Apsley.

A good example of the Braunston high style of decoration as a background to life. Considerable visiting took place when waiting to load, as the collection of cups here shows.

Rene Carter has a moment of leisure with a magazine on the long counter of motor-boat *Nelson*. Note the brass capping stud to the tank below, seen fuelling up on page 95.

Another domestic scene showing more of the labour-intensive but effective washing devices. Mrs Taylor (of 'Four Boat Joey' fame) washes by hand in comparative comfort while dolly tub, dolly and mangle were used for the line of heavy washing behind.

Mrs Mary Berrill, who came from Nuneaton, cleans the cabin end of the British Waterways butty *Hale* while waiting for the lock.

In the middle of the locks at Hillmorton, Rugby, passing the Oxford canal maintenance yard arm. Sarah Hambridge has had time, in the three-hour pound below the locks, to wash all the soft furnishings for the cabin and more besides.

This cheerful group, with the gallon cider jar in its wicker case between them, is on Charlie MacDonald's British Waterways boat. Phyllis, with Joycey, looks on tolerantly and Bill Anderson and Sidney MacDonald flank Charlie.

OTHER PEOPLE, OTHER PLACES

In the following group I have tried to show that Robert Longden's interest in the life on the canals extended beyond the record that his many visits to Suttons could achieve, rich and varied though that was. He took time and trouble to pursue his passionate interest elsewhere and, even though there are far fewer such photographs, he took the sort of pictures which could be drawn into sequential meaning. He left no notes whatsoever and no lists or titles to any of his work. The slides, when found, were arranged entirely at random and provided no clues.

Inevitably there are some which seem to be supplementary, where I have described what is seen or endeavoured to provide an explanation. Some still remain impenetrable and the research of others may eventually clarify what is happening or the identity of the people in them.

As I looked at these pictures it seemed that fundamental change was being revealed. The vision and sense which we have of the boat people from all the other pictures, and the cohesion of the community, is being encroached upon. Even without hindsight some people were aware that the life as they knew it would most probably be extinguished. It was not hard to realise that it must be so. The 1944 Education Act required an annual number of school attendances which could not be met by the limited number of days at turn-round points or other stops, when the children could go to school. The canal carriers and the agents concerned with long-distance and family boats were aware that the problem was intractable. At the same time many regular consignees in the coal trade began to look at costing and wake up to the possibilities of alternative means of supply, with the result that it would become much harder to obtain contracts. Boat people had begun to drift off to cottages. Some children had already gone, causing anguish and uncertainty to some parents, to the boarding hostel at Erdington, Birmingham, which had been set up to receive them and to ensure compliance with the Education Act.

The change was slow and the life only gradually diminished. To it, in increasing numbers, came people from the land to cruise the waterways as leisure time increased and, as the influence of the Inland Waterways Association strengthened, some new businesses, besides more hotel boats, followed. Among the literature of the canal, two writers in particular seem to me to cover this transitional time most effectively. One is the boatman Tom Foxon, with two books, *Anderton for Orders* and *'No. 1'*. This latter is a great tale about starting to trade at a late stage and achieving results by pragmatic ability to seize opportunities and be in the right place at the right time. The other writer is David Blagrove and his book *Bread Upon the Waters* (all three are listed in the Bibliography at the end) is both very funny, graphically descriptive of all the working practices, and full of the same people and places as the photographs here. I greatly admire this book where the events are described so entertainingly, and at the same time the writer's historical sense can place these happenings in a much wider context. Some indication of these changes is present in these few books and in the miscellaneous groupings of images which follow. One of the most powerful was taken in Hillmorton locks and the Braunston pictures, for example, are masterly. Whenever I look at these I seem once more to be there.

An exceptional happening rather than another scene. It may be a stoppage since most of the protagonists are boaters, not maintenance men. A pastime when held up in much-frequented tie-ups was to see what could be found with rakes along the bottom of the canal. Here the lock is empty and something is wrong. The Stokes family can be seen, Tom Peasland and Ernie Bray, and Sarah Anne Carter, to the left in the foreground.

On another occasion it is clearly a dropped gate and the maintenance lengthmen are out in force to rectify it. They have the lock gate raised by an anchorage for the ballance beam in the lockside masonry and a pull to a peg in the lock's side. Frank Wenlock and George Smith snr look on and a loaded day-boat waits to get to 'the Light'.

Two Oxford canal lengthmen in the Hillmorton pound clear back the rich summer growth with scythe and rake, not for cosmetic purposes but to facilitate bank maintenance.

A typical scene of departure from the coal fields. A British Waterways butty *Bakewell* is loaded, shipshape, and pulled on a long tow towards Hillmorton locks. It is painted in the new nationalized livery of Blue and Yellow. A good line of washing blows free and more coal dust has been washed out of garments hanging on the tiller.

The slog of a hard pull at the bottom of Hillmorton locks near Rugby. The scour from the emptying lock made 'the bottom too near the top' along the tow-path where the pull had to be made. It looks as if two pairs are working 'butty' — that is, friends helping each other. Both boats in the picture are butties and the motors are already in or entering the paired locks out of sight.

At the bottom lock of Hillmorton Three the Jackson family 'draw a ground paddle' – that is, one not in the gate itself but letting the lock water down through a brick culvert (as visible in the picture on page 116).

Boats *Cairo* and *Warwick* with George and Sonia Smith at Bedworth Bridge, waiting to load up with coal from Newdigate Colliery for Messrs Kearley and Tongue. Their visitors are Michael and Polly Rogers and Michael's brother, Peter. The Rogers had recently bought *Mabel* (from Mrs Mabel Wilson) to work their 'Number One' carrying coal.

A rear view of *Mabel* setting out with a load of house coal for Messrs Atkins of Banbury. Michael Rogers is steering.

Mabel as a newly converted hotel boat. The conversion was carried out at Tooley's Yard in Banbury, much of the work being done by Michael himself. The converted butty *Forget-me-Not*, which draws along the tow-path side, is working with her as a pair. Michael and Pat Streat's two boats *Nancy* and *Nelson* were the first in this enterprising new form of canal life, soon joined by the Rogers' boats.

Before *Mabel* was converted she joined the Smiths with their motor-boat *Halifax* and butty *Warwick*, to run visitor trips out from the basin at Market Harborough to Foxton Locks during the 1950 Inland Waterways Association Festival. Health and Safety officials would blench at such arrangements now, but there were no disasters I heard of during any of the early passenger trips.

In other ways the working life received outside attention. BBC producer Eileen Molony and her husband Peter taste the life on *Warwick*, here seen entering the Oxford lock at Suttons. Eileen Molony was making a programme for the BBC to expose the distresses and opposition to the blanket uniformity of the Blue and Yellow livery introduced by the nationalized British Waterways. The programme, *Roses and Castles*, with the boaters' voices and those of London aesthetes mingled, went out in July 1949.

A friendly non-professional demonstrates how not to get maximum purchase on a tow. Peter Molony, Sonia and George Smith haul loaded butty *Warwick* into the Oxford lock.

Cairo and *Warwick*, with George and Sonia Smith and the Molonys on board, line up in the basin for their turn in the Oxford lock.

The photographer above on the Britannia Bridge catches the inspirer of so many journeys, conversions and enthusiasms. Tom Rolt, author of *Narrow Boat*, at the stern of *Cressy*, on the way home to Banbury from Atherstone in the autumn of 1949.

A shot taken on the same occasion. *Cressy* is just passing from the Coventry canal under the Britannia Bridge.

Cressy in the Oxford lock. Angela Rolt and a young local helper wait to shut the bottom gate of this shallow lock. These stern pictures show the unmistakeable 'Shroppie' build of the boat and the traditional deeply cut name on the stern plank.

The first Waterways Festival at Market Harborough in 1950. Pleasure cruisers, conversions and working pairs congregated, with narrow boats only allowed in the basin, and received an astonishing national press coverage. Working Ovaltine boats *Mimas* and *Rae*, with the Grand Union pair skippered by Mr and Mrs Alf Best, earned constant attention and visits.

Major and Mrs Fielding of the Salvation Army on their converted butty, now named *Salvo*, stop to take on water by the bridge. Their daughter, Jean Fielding, home from school, stands on the tow-path. Mrs Fielding kept up a standard to match any boatwoman's best.

Another view of the fore-end of the Salvation Army boat *Salvo*. The Fieldings normally lay moored along the tow-path on the Coventry side of Hawkesbury Junction. They always left room for other boats to moor temporarily to pick up orders or on occasion to lie there for a maternity case or illness. The Fieldings had originally been at Fenny Stratford. I would suppose friendship and more contact with the community would have been increased when they moved to this much busier scene.

Samuel Barlow Coal Company butty-boat *Warwick*, loaded, but not looking particularly shipshape, lies alongside the old stables in Braunston boatyard. This is the first of a series of pictures taken by Robert Longden in which he follows the working of a pair of boats up a flight of locks.

From the tow-path near Butcher's Bridge at Braunston, *Warwick* moves towards the bottom lock of the flight. Large side-boards are used so that the coal could be piled up at the stern-end for an even trim, and to maximize the load.

The butty has been loosed off to run into the lock and is now checked at the first of the two bollards. The motor is idling in while the steerer has stepped off to close the gate behind it. He will be in time at the head of the lock to draw a paddle which will admit water to slow it down. The dry dock is at the side of the lock.

Left: The butty steerer flicks the hemp strap over the gate, checks the boat at the first bollard and then ties her up to the second and shuts the gate.

Right: The two boats rise together in the lock. The motor will move forward as the level is made with the water in the pound above, and her steerer will pick up the tow as he passes the fore-end of the butty.

At the top lock of the Braunston flight the boats (with headlamp) move out towards Braunston tunnel.

Any visit to Braunston boatyard would include a trip to Ike Merchant (ex- 'Number One') in his sailmaker's shop. For the camera he shows a magnificent cotton line fender for a motor's stern-end. On his shelves jostles other equipment: paraffin headlamps, tun dish (funnels) and tarpaulin top-cloths. These only earned replacement after a given period of time.

Ike is at genuine work on a hemp tip-cat, the elongated fenders which fitted at the back of the motor. Two of these and the fender beyond buffered any necessary stop and protected propeller 'blades' and 'elum' (helm) in the water below. Sets of new ropes were also earned in the fullness of time, but Ike was open to persuasion and had one or two favourites.

The paint shop contained the most beguiling and hypnotic of all boat-docking activities. The one and only Frank Nurser shows off a hand bowl and cans arranged above massive rope coils. From these the right lengths for different needs will be cut and spliced.

The young Ron Hough, who took full advantage of these 'prentice days to become a master painter, decorates an army billy-can which became an unorthodox coal-bucket.

A sight not to be repeated. A new boat and a new oak stern-post under the sheds at Braunston, the adze which has largely formed her boldly set across her bows. Charlie Nurser, Frank's brother, who with him had owned the boatyard before Samuel Barlow Company took it over, would have come down from the village and by a combination of familiar signs aloft and runic heel markings in the earth below, her keel would have been laid. The boat is probably the motor *Ian* and by such traditional measurings, would every wooden narrow boat have been created.

A lifetime in the skills of traditional boat-building: Jesse Owen holds his caulking hammer under his arm and pulls from the coil of oakum to put a strand through the grease pot before filling any crack in the hull. The sound of this softened thudding would be recognizable anywhere. The boat is new; perhaps Longden returned to see *Ian* throughout her making.

Interviewers descend on a new hotel boat, any publicity for an infant enterprise being useful. Henry Monk, who is speaking, was a member of a famous boating family and had worked as crew on *Mabel*.

In 1951 Longden went to the Festival of Britain. No one now knows what interested him there, but this is surely a significant shot. Dwarfed beside a Thames barge, dressed overall, lie a pair of representative Grand Union boats from British Waterways, got up in patriotic red, white and blue for the occasion.

LAST IMAGES

It would be hard to say why these particular images are now assembled as 'last'. But as far as this book is concerned, there has to be an end. I have gathered together here a very few of those pictures I had remembered best as being most characteristic of the photographer's work or epitomizing the core of the life itself in passing beyond anything he might himself have thought possible.

This group scarcely needs captions because of the strength of the images, but I shall follow the usual practice and say of each what I know of them. There is not a particularly cheerful one among them, the merriment which is seen here and there earlier in this book is absent and that is wrong; it should be here. There is a wonderful picture of lovingness; a supreme portrait of pride, fortitude and endurance; an image of superhuman physical endeavour which is really quite difficult to look at and there are those things which the photographer thought beautiful or deemed to be meaningful. Perhaps his own reticent and withdrawn spirit, as observed and recorded, lies in them.

The last in the book shows one aspect of the scene where all that has gone before has taken place. The boats have mostly moved, the people gone. The trodden snow shows the degree of vigorous life which has passed upon it, but with the thaw those evidences of the life will be gone too.

A second study of hands and ropes, with the firm arms of Sarah Stokes and a cotton line worn enough to become tractable in her hands. The angle of the rope means she is pulling from the mast of the Ovaltine butty *Rae*.

This bollard became the star of many photographs besides this one by Robert Longden. It survived for a considerable time afterwards and was eventually given an honourable home nearby where it languished unused and finally decayed.

A husband and wife team, probably without a mate or help, work a pair of loaded boats up the three locks at Hillmorton near Rugby. The two single locks for each lift side by side and the silted up canal below, made working them a mixture of effort and ingenuity. Here, for the wife, effort seems too weak a word.

Conditions were always very hard when the children were small. Sam Higgins is away in this affecting picture and Mary, his wife, is doing her best with two families to care for; two of her own and four others. Only her eldest stepson, left, is big enough to help with the boats and the immediate problem will be solved, sensitively or insensitively, by local social workers.

Ada Boswell shows what hard labour was involved in harsh conditions. She worked two-handed with her husband, Tommy, a very small man, and was the official steerer for the pair. Here she hauls the loaded butty to the Oxford lock.

A Samuel Barlow Coal Company butty *Montgomery* from Braunston is tied up awaiting orders. In the pause from work, Fred Rice cares for his infant son.

I have not found anyone who can identify the stoical boatwoman in this fine portrait. Perhaps she was a 'northerner' or from one of the Fellows, Morton & Clayton pairs taken over by British Waterways, who only then began to frequent Hawkesbury. She wears a traditional boatwoman's bonnet, graceful and less flamboyant than many. These were already rare and Robert Longden always tried to record both bonnet and wearer.

The Coventry canal at Hawkesbury, looking towards the single track railway bridge. The line ran towards 'Coventry Light' from Longford sidings and had also served abandoned collieries beyond. The canal curves away to Longford and the gasworks.

FURTHER READING

A literature of the life of the working canal population barely existed before the last war. *Rob Rat* and *Our Canal Population* (both 1878), the latter by the reformer George Smith, were both polemical efforts which tried to bring people's attention to the boating community and to what was thought to be the intolerable conditions in which they lived. Legislation for the registration and inspection of canal boats for domestic occupation and, in theory at least, to limit numbers in the cabins and ensure standards of maintenance, was what rewarded Smith's efforts. The long jump to the publication of E. Temple Thurston's *The Flower of Gloucester* in 1911 and A.P. Herbert's *The Water Gipsies* in 1930, barely troubled the surface of these unnoticed lives.

With the war and the introduction of women trainees to the canals or 'cut', this was to change and the publication of books on the subject, presented in various ways, began. This has included several remarkable collections of photographs gathered by Michael E. Ware, Hugh McKnight and others. It does not look likely to stop at the time of writing, as more people take to the water for pleasure and the associations of historic working boats in museums or in private hands strengthens. These historic boats have an aura about them while they survive, of the life that was so little regarded.

I list here just those books I have myself or know which could interest or give an insight to anyone who may wish to follow up the subject. I place them in chronological order. Some are mentioned in the text, but the list is not definitive and does not mention one important primary source. This consists of the various government papers or bills which were introduced as a result of commissions called to investigate the 'living in on canal boats', where those with the boat people's interests at heart gave evidence or, in some instances, the boatmen themselves spoke.

1878 Pearse, Mark Guy. *Rob Rat*

1878 Smith, George. *Our Canal Population*

1911 Thurston, E. Temple. *The Flower of Gloucester*

1930 Herbert, A.P. *The Water Gipsies*

1944 Rolt, L.T.C. *Narrow Boat*

1947 Woolfitt, Susan. *Idle Women*

1948 Smith, Emma. *Maidens' Trip*

1950 Rolt, L.T.C. *The Inland Waterways of England*

1950 Maré, Eric de. *The Canals of England*

1962 Boucher, Cyril T.C. *The Pumping Station at Hawkesbury Junction* (Newcomen Society Paper)

1963 Rolt, L.T.C. *Thomas Newcomen. The Prehistory of the Steam Engine* (revised and enlarged by J. S. Allen, 1977)

1965 Wilkinson, Tim. *Hold on a Minute*

1966 Hadfield, Charles. *The Canals of the East Midlands*

1966 Pevsner, Nikolaus and Wedgwood, Alexandra. *The Buildings of England – Warwickshire*

1973 Gayford, Eily. *The Amateur Boatwomen*

1974 Lewery, A.J. *Narrow Boat Painting*

1975 Hanson, Harry. *The Canal Boatmen 1760–1914*

1984 Blagrove, David. *Bread Upon the Waters*

1987 Cornish, Margaret. *Troubled Waters*

1991 Foxon, Tom. *'No. 1'*

1993 Stewart, Sheila. *Ramlin Rose*

1996 Tony Lewery. *Flowers Afloat*

ACKNOWLEDGEMENTS

There are so many people to thank for their help or advice with this book that I cannot place the weight of their deserved gratitude in any order. I hope it is enough to say that without any one of them, the book would have been much less than it is.

My unbounded thanks and gratitude, therefore, goes out to the following: Katherine Adams, J.S. Allen, Mark Baldwin, Cyril Boucher, Laura Carter, Tony Conder, Bert Dunkley, Julia Elton, Dr Peter Foss, Tom Foxon, Jenny Glynn, the late Charles Hadfield, Jim Jackson, Roy Jamieson, Hugh Jones, Sir John Knill, Tony Lewery, John Lewis, Ada Littlemore, Ian Mackenzie Kerr, Hugh McKnight, Michael and Joyce Rogers, Tim Rolt, Sheila Stewart, Paul Walker, Ben Weinreb, Mike West, Peter White, Rose and Bill Whitlock, Tim Wilkinson, Mary Wilson, Ron Wilson, and Brenda Wythey. For this edition, gratitude for changes and extensions of identification in some instances to John and Phyllis Saxon of Bedworth, Noel and Rene James of Hillmorton, and Tony Miles.

Last but not least, I must thank the members of Robert Longden's family who displayed such patience and gave unstinting help where they could. I would particularly like to thank the late Robert Longden, the photographer's son, and his son, Michael.

Finally my thanks go to George Smith of the boats *Cairo* and *Warwick* as shown in the photographs.

Sonia Rolt